Webs

Julie Murray

Abdo Kids Junior
is an Imprint of Abdo Kids
abdobooks.com

Abdo
ANIMAL HOMES
Kids

abdobooks.com

Published by Abdo Kids, a division of ABDO, P.O. Box 398166, Minneapolis, Minnesota 55439.

Printed in the United States of America, North Mankato, Minnesota.

052019

092019

THIS BOOK CONTAINS RECYCLED MATERIALS

Photo Credits: iStock, Shutterstock, ©Brocken Inaglory p.19/CC BY-SA 3.0

Production Contributors: Teddy Borth, Jennie Forsberg, Grace Hansen

Design Contributors: Christina Doffing, Candice Keimig, Dorothy Toth

Library of Congress Control Number: 2018963328

Publisher's Cataloging-in-Publication Data

Names: Murray, Julie, author.

Title: Webs / by Julie Murray.

Description: Minneapolis, Minnesota : Abdo Kids, 2020 | Series: Animal homes | Includes online resources and index.

Identifiers: ISBN 9781532185267 (lib. bdg.) | ISBN 9781644941232 (pbk.) | ISBN 9781532186240 (ebook) | ISBN 9781532186738 (Read-to-me ebook)

Subjects: LCSH: Animal housing--Juvenile literature. | Spider webs--Juvenile literature. | Spider web animals--Juvenile literature. | Animals--Habitations--Juvenile literature.

Classification: DDC 591.564--dc23

Table of Contents

Webs

A web is made of **silk**.

It is sticky. It is strong.

Most spiders spin webs.

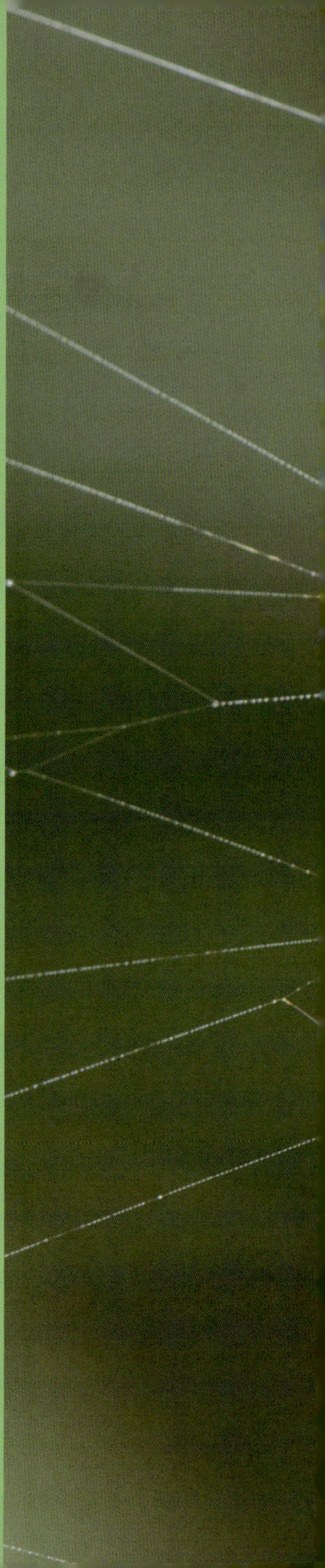

A web traps **prey**. The fly is stuck in the web.

An **orb web** is round.

A sheet web is made on plants.

A tangled web is messy.

A **funnel web** is on the ground.

Tent caterpillars make webs too.

The web keeps them safe.

Have you seen a web?

What Lives in a Web?

black widow spiders

grass spiders

gypsy moths (caterpillars)

yellow garden spiders

Glossary

funnel web
a web with a shape that has a hollow opening to crawl through.

orb web
a web that is shaped like a circle.

prey
an animal or insect that is hunted by another for food.

silk
a fine, soft, shiny fiber made by certain insects.

Index